F·I·R·E

Art Director: Rita Marshall
Book Design: Stephanie Blumenthal
Text Adapted and Edited from the French language by Kitty Benedict
Library of Congress Cataloging-in-Publication Data
Benedict, Kitty.
Fire/written by Andrienne Soutter-Perrot; adapted for the American reader
by Kitty Benedict; illustrated by Patrick Gaudard.
Summary: Introduces the properties and uses of fire.
ISBN 1-56846-043-0
1. Fire—Juvenile literature. 2. Combustion—Juvenile literature.
[1. Fire.]
I. Soutter-Perrot, Andrienne. II. Gaudard, Patrick, ill. III. Title.
QD516.B418   1992
541.3'61--dc20   92-5978

FIRE

WRITTEN BY

ANDRIENNE SOUTTER-PERROT

ILLUSTRATED BY

PATRICK GAUDARD

CREATIVE EDITIONS

Fire seems almost like a living thing. It needs air and food. It is hot and always moving.

Fire has no shape, for it is always changing. It changes color, too. Sometimes it is yellow or orange, sometimes red, and sometimes green or blue.

Fire weighs nothing. You cannot hold it in your hand.

What *is* fire, then? Fire is the light and heat that is produced as things burn.

HOW IS FIRE MADE?

A fire can be started with kindling, wood, and oxygen. But first it must be lighted by a tiny bit of intense heat—a spark.

Kindling and firewood can be found anywhere in the forest. The air is full of oxygen. But where does a spark come from?

You can make a spark by hitting two stones together very hard, or by rubbing two dry sticks together.

Using a match is even simpler. The flame of the match lights the
kindling, and the kindling lights the pile of firewood.

Once the wood catches fire, it burns on its own. Flames dance and embers glow. The fire gives off heat and light.

As the fire burns, it also makes a gas. This gas, mixed with fine ash, makes smoke. It contains water vapor and carbon dioxide.

If the fire is not fed, the flames will die down. The embers will stop glowing, and the fire will go out.

You can also put out a fire by spraying water on it, or by covering it with soil to cut off the air supply.

A little heap of ashes is all that is left of this cooking fire.

HOW IS FIRE USED?

When wood burns, it turns into smoke and ash, giving off heat and light in the process.

People use the heat from fire to cook their food and warm their houses.

Fire is a tremendous source of useful energy. It can be used to melt metal, run engines, or make electricity.

Combustion is another name for burning. In this process, oxygen turns wood or other fuels into energy and waste products.

The combustion of coal, gas, or oil produces a great deal of energy in a short period of time.

A different kind of combustion takes place in our bodies, when
the food we eat is slowly burned by the oxygen we breathe in.

Day and night, your body uses the energy from this slow combustion to stay healthy.

When you play or work, you need extra energy. Your body burns
more food and you feel hungry.

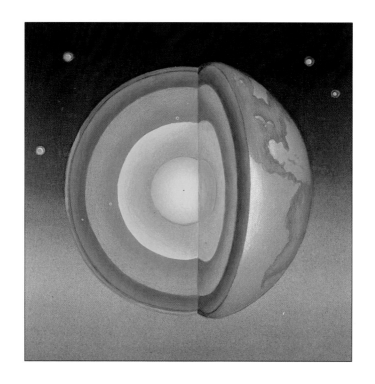

WHERE IS FIRE FOUND?

The Earth is a huge ball covered with a thin layer of rocks and water. This layer is called its crust.

Beneath this crust, the rock is so hot and under so much pressure that it is liquid.

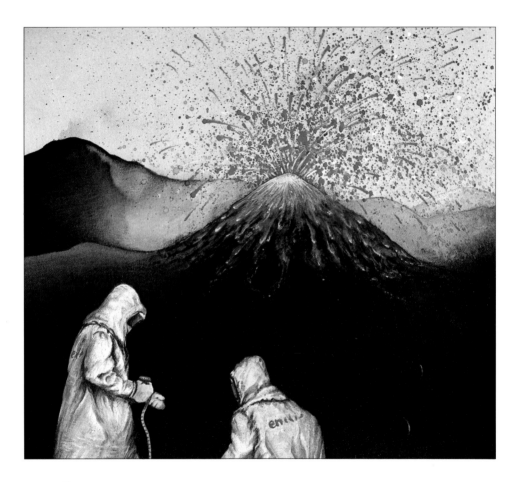

Sometimes the molten rock bursts through a crack in the Earth's crust. This is called a volcano. The gases in the molten rock catch fire when exposed to the air.

The Earth revolves around the Sun, a ball of fire thousands of times bigger than the Earth.

For billions of years, the Sun has been warming our planet.

Plants use the Sun's energy to grow and to produce oxygen. We need plants in order to breathe, and to use for food and shelter.

Without the Sun, without energy, without fire, there would be no life on Earth.